Attainment's EXPLORE budgeting

Judi Kinney

Explore Budgeting

By Judi Kinney

Tom Kinney, Editor
Jo Reynolds, Graphic Design

An Attainment Company Publication

Printed in the United States of America
ISBN 1-57861-819-3

Attainment Company, Inc.

P.O. Box 930160 • Verona, Wisconsin 53593-0160 USA
Phone: 800-327-4269 • Fax: 800.942.3865
www.AttainmentCompany.com

Table of Contents

Chapter 1

How Much Does It Cost?

Vocabulary

1
discount

to buy something at a lower price

2
online

using the Internet

3
receipt

shows how much something costs

4
sale

to sell at a lower price

5
sales tax

a tax, or money added to the price of something

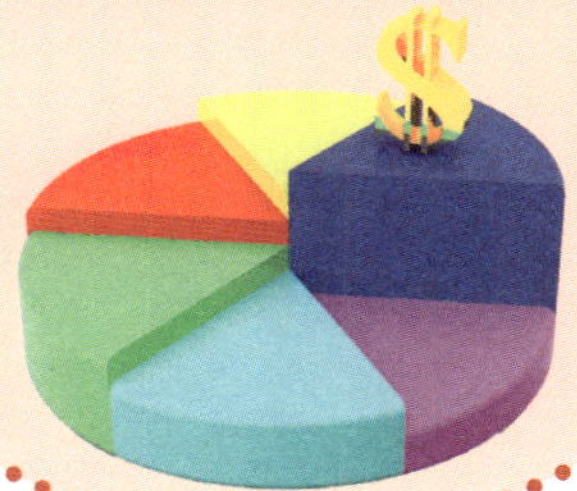

6
tip

extra money given for good service

The Coffee Shoppe

	Small	Medium	Large
Regular Coffee	$1.60	$1.80	$2.00
Coffee Latte	$2.95	$3.35	$3.75
Cappuccino	$3.00	$3.50	$4.00
Espresso	$1.65	$1.95	$2.25
Coffee Mocha	$3.55	$4.25	$4.95
Regular Tea	$1.95	$2.10	$2.40
Chai Latte	$2.50	$2.95	$3.35
Chocolate	$1.60	$1.80	$2.00

On her way to work each day Jessica stops to buy a coffee drink at a local shop. She budgets or sets aside $4.00 a day to buy her drink. Today Jessica wants a medium cappuccino. She takes out her wallet to count her money before she pays for the cappuccino.

Money in Jessica's Wallet

Answer

Does Jessica have enough money to buy a medium cappuccino?

a. yes   b. no

Answer

1. How much money does Jessica have left after she buys a cappuccino?

   a. $.25   b. $.50   c. $.10

2. Jessica changes her mind and wants to buy a large coffee mocha. Does she have enough money?

   a. yes   b. no

3. What would you buy if you had $4.00?

   ____

The Coffee Shoppe, Part 2

Jessica has been saving the change from her daily purchases at The Coffee Shoppe. Today she got up late and didn't have breakfast. Jessica decides to buy a doughnut to go along with her coffee drink. She counts her change to see if she has enough money for a doughnut.

Answer

Does Jessica have enough money for a doughnut?

a. yes

b. no

Answer

1. How much money does Jessica have in her wallet?

   a. \$.25 b. \$.85 c. \$.80

2. Jessica finds two quarters at the bottom of her purse. She has \$1.35 to spend. What can Jessica buy instead of a doughnut?

   a. cookie b. Danish c. muffin

3. If Jessica saves \$.50 a day for 3 days, how much money will she have to spend on a pastry? (show work)

Discounts

Heilke and Patrice plan to watch a movie that Patrice downloaded onto her TV.

Heilke will buy a pizza. She has $20.00 left that she has budgeted for entertainment. Heilke printed out a coupon from the Internet. The coupon gives Heilke a discount. A discount lets Heilke buy the pizza at a lower price. Heilke's favorite pizza costs $15.00.

Answer

How much is the discount on this coupon?

a. 10%   b. 100%   c. 1%

Answer

1. What is the regular price of Heilke's favorite pizza?

   a. $5.00   b. $15.00   c. $13.00

2. How much money will Heilke save using her discount? (show work)

3. With the discount how much did the pizza cost? (show work)

Downloading Music

Benton wants to download music to his portable player. He went on the Internet to find the cheapest company. Benton found one that charges $.89 a song and he got the first 5 songs free. Happy, he downloaded the 5 free songs. He paid for 10 more.

Answer

How many free songs does a new customer get? ____________

Answer

1. How much did Benton pay for each song?

   a. $.98 b. $.89 c. $.99

2. How many songs did Benton buy?

   a. 5 songs b. 15 songs c. 10 songs

3. How many songs did Benton download all together?

   a. 5 songs b. 15 songs c. 10 songs

Movie Tickets

Patrice is going to a matinee movie with her friend Sara. A matinee ticket costs less than a regular ticket. Each ticket is $10.00. Patrice decides to bring $20.00 so she can buy popcorn and a soda. When Patrice gets to the theater, she discovers she only has $15.00! What can Patrice buy to eat?

$5.50

$3.75

$4.60

$2.80

$3.00

$1.75

Answer

Write one thing that costs less than $5.00. ______________________

Answer

1. How much does a movie ticket cost?

   a. $1.00 b. $20.00 c. $10.00

2. If Patrice paid for a ticket at regular prices it would cost $15.00.
   How much money did she save by buying a matinee ticket? (show work)

3. On page 16, ( ) one refreshment Patrice can buy.

Jeans

José needs new jeans. He shops at his favorite jean store. He has been saving money to buy them. So far José has saved $35.00. The jeans cost $49.50 plus tax. He was mailed a flyer announcing a sale on jeans. On sale, the jeans cost $29.95. José buys a pair.

Answer

How long is the sale?

a. a week   b. a weekend   c. a day

Answer

1. How much were the jeans before the sale?

   a. $29.95   b. $35.00   c. $49.50

2. How much were the jeans on sale?

   a. $29.95   b. $35.00   c. $49.50

3. How much did José save? (show work)

Online Special

José saved $140.00 to buy a new pair of tennis shoes. He saw a pair at the mall for $132.00. He also has to pay 5.5% sales tax. José found the same pair online for $10.00 less. **Online** means he found it on the Internet. That means he wouldn't have to pay sales tax or shipping charges. José buys the shoes online.

Answer

Does José have to pay shipping costs?

a. yes   b. no

Answer

1. Not including tax, how much would the shoes cost if José bought them at the local mall?

   a. $140.00   b. $132.00   c. $10.00

2. How much did the shoes cost José when he bought them online?

   a. $140.00   b. $132.00   c. $122.00

3. What was the difference between the prices of the tennis shoes?

   a. $18.00   b. $122.00   c. $10.00

Video Games

Heilke and her friends love to play video games. She saved $40.00 to buy a new game. Heilke went to a big box store. She found a game she liked for $35.95. She also had to pay sales tax. Sales tax is extra money added to the price of things. Using her cell phone, Heilke calculated the total price.

BIG BOX STORE
Anytown, USA 55350

STORE 00000 REGISTER 0000
CASHIER JOE B DATE 4/8/14

Sales Receipt

ITEM	PRICE
VIDEO GAME	$35.95
TAX	$2.16
TOTAL	$38.11

PAYMENT CASH

Answer

What was the total price of the game?

a. $40.00　　b. 35.95　　c. $38.11

Answer

1. What was the price of the game?

   a. $35.95　　b. $2.16　　c. $38.11

2. What was the amount of sales tax Heilke paid?

   a. $35.95　　b. $2.16　　c. $38.11

3. on the sales receipt how much Heilke paid for the game.

A Smart Phone

Dana works as a marketer for a company. She wants a smart phone. It would make her work easier. Dana used the Internet and found a nearby store that sells the phone she wants. The phone cost $199.00 plus 6% sales tax. Dana has $220 she can take out of her savings account to buy it.

Answer

How much did Dana withdraw from her savings account?

a. $220.00 b. $22.00 c. $2.00

Answer

1. What did Dana use to find the store that sold her smart phone?

   a. phone book b. magazine c. Internet

2. Where did Dana get the money to buy her phone?

   a. checking account b. savings account c. credit card

3. Including tax does Dana have enough money to pay for her phone?

   a. yes b. no

Eating Out

José often eats in restaurants. He really likes to eat in one small sidewalk café. He tries to keep his meals under $20.00 including a tip. A tip is extra money given for good service. Today José orders a meal that costs $12.95, ice tea that is $2.10, and he pays $.65 tax. Can he tip the server $3.00?

Answer

How much does José budget each time he eats out?

a. $2.00 b. $10.00 c. $20.00

Answer

1. How much was José's meal?

   a. $12.95 b. $2.10 c. $.65

2. How much was José's total bill? (show work)

3. Does José have enough money left to pay the server a $3.00 tip?

   a. yes b. no

Taking a Friend Out to Eat

José's best friend just got a new job. He is taking his friend to his favorite restaurant to celebrate. José knows he needs to bring more than $20.00 to pay for two people. He decides to take $45.00. After the meal, the server gives José a receipt. The **receipt** shows how much his meal cost.

Anytown Sidewalk Café
1335 Main St., Anytown, USA 55350

REGISTER 0000 SERVER MARY
DATE 4/8/14

Receipt

ITEM	PRICE
MEAL	$13.55
MEAL	$14.50
COFFEE	$1.00
ESPRESSO	$1.82
SUBTOTAL	$30.87
TAX	$1.63
TOTAL	$32.50
TIP	$6.50

Answer

Did José budget enough money to pay for the meal?

a. yes   b. no

Answer

1. Subtotal means the amount the meals cost before tax was added. What was the subtotal for José's bill?

   a. $1.63   b. $30.87   c. $32.50

2. How much did José spend for the meal including the tip?

   a. $30.87   b. $32.50   c. $39.00

3. How much money did José have left? (show work)

Don't Forget Sales Tax

Benton has budgeted $25.00 to buy a T-shirt of his favorite NCAA team. Benton finds one that he likes for $25.00 at the local sports store. He remembers that his state has a 6% sales tax. He uses the calculator on his phone to figure out how much he has to pay for the shirt, including the sales tax.

$32.00

$22.00

$25.00

$18.00

Answer

How much money has Benton budgeted to buy a T-shirt?

a. $2.50 b. $25.00 c. $52.00

Answer

1. Benton has to pay a 6% sales tax. How much tax does he have to pay on a T-shirt that costs $25.00?

$$\begin{array}{r} \$25.00 \\ \times\ .06 \\ \hline \square \end{array}$$

2. Including tax, does Benton have enough money to buy the T-shirt?

a. yes b. no

3. ◯ another T-shirt that Benton can buy. Remember to include sales tax.

A New TV

Patrice needed a new TV. Her local appliance store was having a sale. A **sale** meant she could buy a TV at a lower price. She found a TV she liked. Patrice budgeted $900 to buy a TV and had it mounted on her wall. She also had to pay 5% sales tax.

Sales Receipt

Appliance Store

1 Mall Court Anytown, USA 53778

DATE 4/8/14

ITEM	PRICE
TV	$650.00
WALL MOUNT CHARGE	$200.00
SUBTOTAL	$850.00
SALES TAX	$42.50
TOTAL	$892.50

Answer

How much did Patrice budget to buy a TV?

a. $9.00 b. $90.00 c. $900.00

Answer

1. Did Patrice stay within her budget?

   a. yes b. no

2. How much did the store charge to mount the TV on the wall?

   a. $200.00 b. $650.00 c. $42.50

3. How much money did Patrice have left after she bought her TV? (show work}

Chapter 2

Daily Budgets

Vocabulary

1
budget

a plan for spending and saving money

2
checking account

an account that lets a person deposit and spend money

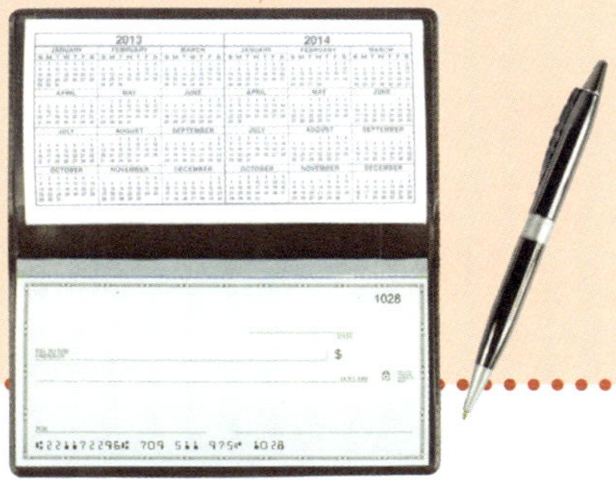

3
debit card

a card that acts like a check

4
deposit

money put into a checking/savings account

5
expenses

money spent on something

6
withdraw

take money out of a bank account

Darren's Day

Darren is a high school student and works part-time. He decided to make a budget for himself. A **budget** is Darren's plan for spending and saving his money. He budgets $10.00 a day to spend. He rides to school on a city bus. A roundtrip ticket costs $4.00. After school he likes to stop at a coffee shop. Today he bought a smoothie for $4.75.

Darren's Expenses

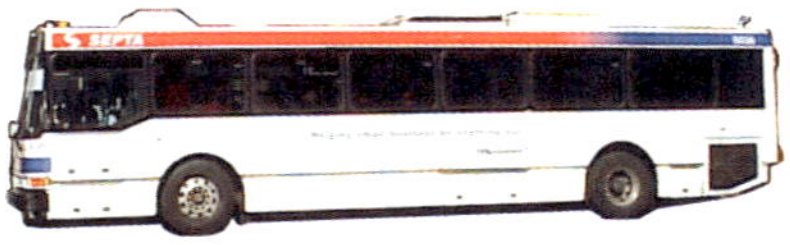

$10.00

-$4.00 to and from school

[ ]

-$4.75 smoothie

Total left [ ]

Answer

Complete the problem on page 38 to find out how much money Darren had left at the end of the day.

Answer

1. Did Darren stay within his budget? (circle one)

   a. yes   b. no

2. How much money does Darren spend on his bus fare each day?

   a. $2.00   b. $4.00   c. $10.00

3. What do you think Darren should do with his left-over money?

   a. save it   b. spend it   c. add it to his daily budget

Cafeteria

Nancy is a nurse and she eats in the hospital cafeteria. She budgets $9.00 a day for her lunch. Nancy usually eats a salad or sandwich and has something to drink. She also has to pay a 5% sales tax. Help Nancy pick a meal. Remember not to go over **$9.00.**

Menu

Salad $6.65 	Sandwich $4.25 
Fruit Bowl $2.85 	Yogurt $1.50 
Coffee $1.50 	Soda $2.00 

Answer

How much money does Nancy budget for lunch?

a. $6.00   b. $9.00   c. $10.00

Answer

1. On the menu on page 40, ◯ a meal Nancy can eat for lunch.

2. Including tax, what is the total cost of the meal? (show work)

3. Is the meal less than $ 9.00?

   a. yes   b. no

A Lawyer

Maria is a lawyer who works downtown. She budgets $35.00 a day for expenses. **Expenses** are things Maria must pay for. She uses a debit card. A **debit card** is like a check. Each time Maria uses it, money is **withdrawn** from her checking account. The bank takes money from her account to pay for things she buys with her debit card.

The Coffee Shoppe
135 Main St., Anytown, USA 53778

Receipt

ITEM	PRICE
COFFEE	$1.80
MUFFIN	$2.50
SUBTOTAL	$4.30
TAX	$.26
TOTAL	$4.56
TIP	$1.00
TOTAL	$5.56

PAYMENT DEBIT CARD

X *Maria Rodriguez*

RODRIGUEZ, MARIA

Anytown Sandwiches
200 First St., Anytown, USA 53778

Receipt

ITEM	PRICE
CHICKEN WRAP	$7.55
ICED TEA	$2.00
SUBTOTAL	$9.55
TAX	$.58
TOTAL	$10.13
TIP	$2.00
TOTAL	$12.13

PAYMENT DEBIT CARD

X *Maria Rodriguez*

RODRIGUEZ, MARIA

Answer

Where did Maria spend the most money?

a. The Coffee Shoppe b. Anytown Sandwiches c. Anytown Parking

Answer

1. How much did Maria pay for parking?

   a. $12.00 b. $5.56 c. $12.13

2. What was the total amount Maria spent or withdrew from her checking account? (show work)

3. Did she stay within her daily budget? a. yes b. no

Anytown Parking

239 Main St., Anytown, USA 53778

Receipt

8 HOURS @ $1.50 AN HOUR

TOTAL $12.00

PAYMENT DEBIT CARD

X *Maria Rodriguez*

RODRIGUEZ, MARIA

A Salesman

Jamal is a salesman for his company. Today he drives out of town for a meeting. His company gave him $65.00 for expenses. Jamal filled his car with gas. The bill was $39.00. He bought a lunch for $18.00. On his way home he bought a cappuccino for $4.50. Jamal counted the money he had left.

Jamal's Expenses

$65.00
-$39.00

Total

-$18.00

Total

-$4.50

Total left

Answer

Complete Jamal's business expenses. Work the problem on page 44.

Answer

1. What was Jamal's largest expense?

   a. cappuccino   b. lunch   c. gas

2. How much money did Jamal have left?

   a. $4.50   b. $3.50   c. $18.00

3. Did Jamal stay within his budget?

   a. yes   b. no

The Lings

Water Park

The Ling family took a day trip to a water park. They withdrew $150.00 from their checking account for expenses. A **checking account** is a bank account where the Lings keep money for expenses. They paid for tickets at the gate. They bought lunch. Later the Lings ate snacks. Afterwards the two Ling children wanted to get T-shirts.

LINGS' BUDGET

$150.00

−$75.00 → Family price

TOTAL [ ]

−$35.00 →

TOTAL [ ]

−$15.00 →

TOTAL [ ]

Answer

Complete the Lings' daily budget to find out how much money they have left. Work the problem on page 46.

Answer

1. How much money does the Ling family have left? ______________

2. Each t-shirt costs $15.00. Do the Lings have enough money to buy two t-shirts?

   a. yes b. no

3. ◯ something else the Lings can buy for the two children.

   a. $64.00 b. $4.00 c. $22.00

A Trip to the Mall

Patrice is going shopping with friends. There is a sale at her favorite store. Patrice made a list of clothes she wants to buy. She has $150.00 to spend. She set aside $12.00 to eat lunch with her friends in the Food Court. Patrice now has $138.00 to spend. She remembers to include 6% sales tax.

Cathy's Clothes

Anytown Mall   Anytown, USA 55350

CASHIER ISABEL   DATE 7/5/15

Sales Receipt

ITEM	PRICE
JEANS	$29.99
TEE SHIRT	$15.00
TEE SHIRT	$12.99
JACKET	$42.99
SUBTOTAL	$100.97
TAX	$6.06
TOTAL	$107.03

PAYMENT CASH

Answer

Did Patrice stay within her budget of $138.00?

a. yes

b. no

Answer

1. How much money did Patrice pay in sales tax?

   a. $100.97

   b. $12.00

   c. $6.06

2. Patrice did not have enough money to buy:

   a. jeans

   b. tennis shoes

   c. a jacket

3. Patrice set aside $138.00 to buy her clothes. How much money does she have left after she bought her clothes? (show work)

Baseball Game

Alejandro and his brother are going to the baseball game. He has budgeted $40.00 for it. Alejandro received a coupon in the mail that cuts the price of his ticket. He gets a 20% discount if he uses it. The ticket usually costs $22.00. The rest of the money he can use for snacks.

Answer

How much is the discount on a ticket to the game?

a. 2%   b. 10%   c. 20%

Answer

1. What is the price of a regular ticket?

   a. $20.00   b. $22.00   c. $40.00

2. How much did Alejandro save using the coupon? (show work)

3. How much did Alejandro's ticket cost after he used the coupon? (show work)

A Daily Savings Plan

When Darren first started working part-time, he didn't have a budget and spent all his money. He had no money for emergencies or to buy something expensive. Darren came up with a daily savings plan. He now puts a dollar a day into a jar and **deposits** the money each week into his savings account.

Saving Change

Answer

How much money does Darren try to save each day?

a. $1.00   b. $10.00   c. $.10

Answer

1. Before trying to save, how much money did Darren spend?

   a. some of it   b. all of it   c. none of it

2. If Darren saves a dollar a day, how much money will he deposit into his savings account at the end of each week?

   a. $1.00   b. $5.00   c. $7.00

3. Darren's parents will match the amount of money he saves. How much money will Darren have at the end of each week?

   a. $2.00   b. $10.00   c. $14.00

Daily Budget Summary

Maria made a budget to keep track of how much she spends each day. It is called a daily budget. A daily budget includes transportation and food for each day. A budget helps keep Maria from spending too much. When she follows her budget she has money to pay for other things.

Maria's Budget

driving to work

parking

buying coffee

eating lunch

Daily Budget

round trip fare $3.00

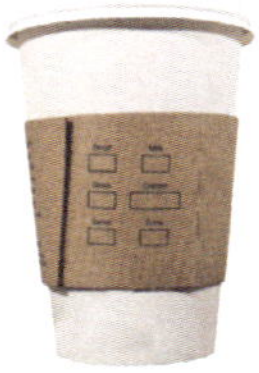

latte $3.35

doughnut $.80

tacos $3.00

pop $1.75

smoothie $2.35

sandwich $4.00

comic $3.99

sale! $6.49

Answer

Your daily budget is $12.00. You must have enough money to ride the bus to and from your house. Using the Daily Budget chart on page 54, ◯ what you can buy after you purchase your ticket.

Answer

1. Add all the things you bought to find the total. (show work)

2. Did you stay within your budget?

   a. yes   b. no

3. Share how you spent your money.

____

____

Another Budget Summary

Ling Family's Budget	

tickets for rides	money for food

Besides budgeting for everyday needs, a daily budget has to be ready for special events, too. The Ling family made a budget for their trip to the water park. They needed enough money for tickets and food for the family. Before you take a day trip, make a budget so there is enough money for your trip.

Daily Budget		

all-day ticket $45.00	**lemonade $3.50**	**shake $4.00**

hot dogs $4.50	**hamburger $6.50**	**veggie burger $7.00**

cap $12.00	**water bottle $8.00**	**stuffed animal $14.00**

Answer

You have $70.00 to spend at the amusement park. You must buy an all-day ticket. Using the Daily Budget chart on page 54, ◯ what you can buy after you purchase your ticket.

Answer

1. Add all the things circled to find the total amount of money spent. (show work)

2. Did you stay within your budget?

   a. yes   b. no

3. Share how you spent your money.

Chapter 3

Weekly Budgets

Vocabulary

1
credit card

a card used to borrow money to buy something on credit

2
deductions

taxes and other things a worker must pay

3
fixed expenses

expenses that stay the same

4
gross pay

total amount of money a worker earns

5
net pay

the amount of money a worker takes home

6
variable expenses

expenses that change from month to month

Benton

Dog Walker

Benton is a high school student who earns money after school walking dogs. He charges $10.00 a walk. He walks two dogs in his neighborhood. Benton earns $100.00 a week. He saves part of his money to go to a technical college when he graduates in two years. The rest of his money he spends on weekly expenses.

Benton's Weekly Budget			Income $100.00		
Fixed Expenses		Amount	Variable Expenses		Amount
	College fund	$40.00		Entertainment	$15.00
				Eating out	$18.00
				Clothes	$12.00
				Misc.	$15.00

Answer

What is Benton's largest expense?

a. college fund	b. entertainment	c. clothes

Answer

1. How much money does Benton put aside each week to save for college?

   a. $100.00	b. $40.00	c. $15.00

2. Benton received $100.00 from a relative for his birthday. Write one way he can budget this money.

   ____

   ____

   ____

   ____

3. Benton buys a sweatshirt for $22.00 and has to spend less money on something else. What do you think he can spend less money on this week?

   a. entertainment	b. eating out	c. misc.

More Money

Benton has another dog to walk. Benton earns $50.00 a week more. He is happy because he now earns $150.00 a week. Benton needs to change his weekly budget. He adds $25.00 a week more to his college fund. He has $25.00 to add to his variable expenses. Help Benton change his budget.

Benton's New Weekly Budget			**Weekly Income $150.00**		
	Fixed Expenses	**Amount**		**Variable Expenses**	**Amount**
	**College fund**	**$65.00**		**Entertainment**	________
				**Eating out**	________
				**Clothes**	________
				**Misc.**	________

Answer

How much is Benton now saving each week for college?

a. $150.00 b. $40.00 c. $65.00

Answer

1. How much more money each week is Benton saving for college?

   a. $25.00 b. $65.00 c. $150.00

2. Using the new amount that Benton now saves, how much will he put aside for college in a year? (show work)

3. Make a new budget for Benton's variable expenses. Fill in the chart on page 64. Be sure not to go over $85.00. Share the budget you made for Benton.

Benton's Budget

Since Benton is still in high school and lives with his parents, he created a budget that lets him have more money for clothes, entertainment, and eating out with his friends. He is saving money for college and puts it into an account. Benton never uses this money for other things.

College Fund Growing

Anytown Bank & Trust

113 S. Main ▪ Anytown, USA 53778

Yearly Statement Account 5678

Total Deposits	Interest	Interest Earned	Total Savings
$3,380	1%	$33.80	$3,413.80

Answer

Saving $65.00 a week, how much did Benton saved for college in one year?

a. $3,380   b. $33.80   c. $3,413.80

Answer

1. By the time he graduates, Benton has saved $6,827.60. His grandparents gave him $200.00 for college. How much does Benton have altogether? (show work)

2. Benton wants to become a vet tech but his first semester tuition is $2,125.15. Has Benton saved enough money to pay tuition for his first semester?

   a. yes   b. no

3. Do you think Benton's budget helped him have enough money to pay for some of his college expenses?

   a. yes   b. no

Greenhouse Worker

Patrice works in a greenhouse. She gets paid every week. Patrice reads her check stub and budgets her money based on the amount she earns each week. Patrice's **gross pay** is the total amount of money she earns. **Deductions** include taxes or health insurance that Patrice must pay. Her **net pay** is the amount of money she takes home.

Ever-Ready Shipment
7250 Oak Avenue
Anytown, USA 53778

Check Number: 9045
Pay Date: 01/25/13

PAY ***Three hundred fifteen dollars and 14 cents*******************$315.14

Pay to the order of Patrice Novak
7998 Elm Street
Anytown USA 53778

Patrice Novak	**Pay Period 1/19/13-1/25/13**	
Earnings	**This Period**	**YTD Amount**
Gross Pay	**$456.00**	**$1824.00**
Deductions		
Federal Income Tax	$68.40	$273.60
State Income Tax	$22.80	$91.20
FICA - Social Security	$27.36	$109.44
FICA - Medicare	$7.30	$29.20
Health insurance	$15.00	$60.00
Net Pay	**$315.14**	**$1260.56**

Answer

How much is Patrice's gross pay?

a. $315.14 b. $456.00 c. $912.00

Answer

1. Gross pay means:

   a. a worker's total pay

   b. things deducted from a paycheck

   c. take home

2. Taxes are part of a worker's:

   a. gross pay b. deductions c. net pay

3. How much is Patrice's net pay?

   a. $456.00 b. $315.14 c. $68.40

Patrice's Budget

Patrice made a weekly budget using her net pay. **Net pay** is the amount of money she takes home. She can't spend more than $315.14. Patrice made two columns: one for fixed expenses and another for variable expenses. **Fixed expenses** stay the same. Patrice pays the same amount for rent. **Variable expenses** change. She doesn't always spend the same amount for groceries.

Patrice's Weekly Budget			
Weekly Income $315.14			
Fixed Expenses	Amount	Variable Expenses	Amount
rent	$110.00	groceries	$60.00
cell phone	$17.50	entertainment	$25.00
heat	$14.00	clothes	$25.00
water	$4.50	misc.	$19.14
Internet	$6.00		
savings	$15.00		
metro bus card	$19.00		

Answer

How much money does Patrice budget each week?

a. $100 b. $315.14 c. $60.00

Answer

1. What is Patrice's largest fixed expense?

   a. rent b. groceries c. cell phone

2. How much money does Patrice try to save each week?

   a. $6.00 b. $15.00 c. $7.25

3. Patrice likes to buy a latte each morning before she goes to work. Where did she budget money to buy a latte?

   a. groceries b. entertainment c. misc.

Shopping for Food

Patrice likes to cook and budgets $60.00 a week for groceries. To stay within her budget, she makes out a grocery list before she goes to the store and sticks to it. She tries to find items on her list that are on sale. Patrice uses coupons to reduce the price of things she wants to buy.

Grocery Store list

- bread
- milk
- cereal
- eggs
- yogurt
- bag of apples
- lettuce
- tomatoes
- chicken
- sandwich meat
- ground beef
- pasta
- bag of frozen beans
- bag of corn
- orange juice
- paper towels

SAVE $1.00

Henry's **Honey Oats**

Limit one per customer Expires 12/31/15

save $.50

Farm Fresh **Frozen CORN**

Limit one per customer Expires 12/31/15

$.50 OFF

***Best Value* PAPER TOWELS**

Limit one per customer Expires 12/31/15

Answer

How much did Patrice save using coupons?

a. $1.00 b. $2.00 c. $2.25

Answer

1. Best Value paper towels cost $1.29. How much did Patrice pay using the coupon? (show work)

2. Why does Patrice use coupons?

____

____

____

____

3. Look at the coupons. When do the coupons expire or no longer can be used?

a. 12/31/15 b. 12/13/15 c. 1/31/15

Miscellaneous Money

Patrice budgets $19.14 to buy such things as lattes, magazines, cosmetics, and lunch with her best friend. She goes to a local restaurant where she can order a meal under $10.00. Today Patrice has $12.00 left from her miscellaneous money to spend on lunch. She remembers the $12.00 must include 6% tax and a $2.00 tip for the server.

Today's Specials

1. Today's Quiche	$6.50
2. Soup of the day (cup)	$4.00
3. Soup of the day (bowl)	$6.00
4. Soup of the day & 1/2 sandwich	$7.50
5. Tuna salad sandwich with chips	$7.00
6. Yogurt and fruit	$5.50
7. Hamburger and fries	$8.50
8. House salad and roll	$7.00
9. Turkey wrap	$7.00
10. Chicken burrito	$8.00
11. Cheeseburger and fries	$9.25
Drinks: ice tea, lemonade, coffee, soda	$1.50

Anytown Sandwiches
200 First St., Anytown, USA 53778

Receipt

ITEM	PRICE
MEAL # ______	______
DRINK ______	______
TAX	X .06% ______
TOTAL	______

Answer

Select a meal and calculate the tax Patrice will need to pay. Use the receipt on page 74.

Answer

1. Add the amount of the tax to the price of the meal. (show work)

2. Is the total amount of the meal under $10.00?

   a. yes   b. no

3. Does Patrice have enough money left to give the server a $2.00 tip?

   a. yes   b. no

Getting a Credit Card

Patrice has a roommate to help with her expenses, but she doesn't always have enough money for emergencies and other expenses. So Patrice got a credit card. A **credit card** lets Patrice borrow money from a bank. The bank pays the bill and charges Patrice extra money for using the card. She now must budget $5.00 a week to pay this bill.

Patrice's Weekly Budget				
Weekly Income $315.14				
Fixed Expenses	Amount	Variable Expenses	Old Amount	New Amount
rent	$110.00	groceries	$60.00	
cell phone	$17.50	entertainment	$25.00	
heat	$14.00	clothes	$25.00	
water	$4.50	misc.	$19.14	
Internet	$6.00	credit card	$5.00	
savings	$15.00			
metro bus card	$19.00			

Answer

How much a week must Patrice budget for her credit card bill?

a. $5.00 b. $15.00 c. $20.00

Answer

1. Why did Patrice get a credit card?

   a. she lives with a roommate

   b. to have money for emergencies

   c. to pay for her variable expenses

2. If Patrice budgets $5.00 a week to pay her credit card bill, how much does she budget each month?

   a. $5.00 b. $10.00 c. $20.00

3. Create a new budget using Patrice's variable expenses so that she can save $5.00 a week to pay her credit card bill. Share your budget.

Oops! Too Many Charges

Patrice must change how she spends her money. Instead of just using her credit card for emergencies, Patrice bought clothes, lattes, meals in restaurants, and movie tickets with it. She has to budget $10.00 a week to pay this bill. To save money Patrice stopped buying lattes on her way to work. She brings a thermos of coffee instead.

No More Lattes

Monday	Tuesday	Wednesday	Thursday	Friday
$3.75	$3.75	$3.75	$3.75	$3.75

Answer

How much does one latte cost? ____________________

Answer

1. Why did Patrice stop buying lattes?

   a. to bring coffee in a thermos

   b. she charged them instead

   c. to pay for her credit card bill

2. How much did Patrice save each week when she stopped buying lattes? (show work)

3. Is Patrice saving enough money each week to pay for her credit card bill?

   a. yes   b. no

Patrice's Budget

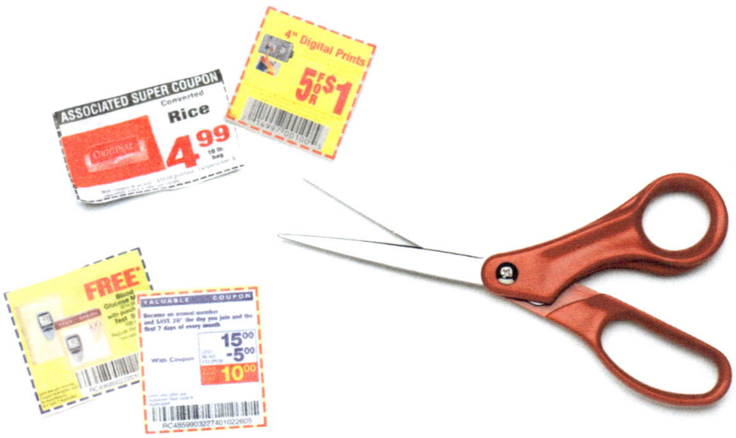

Patrice made a budget using her net pay. She cannot spend more money than she earns. Patrice knows her fixed expenses have to be paid first. When she struggled with sticking to her budget, she decided to change the amount she spent on her variable expenses. Patrice used coupons and spent less on miscellaneous purchases to stay within her budget.

List Ways to Save

1 ______________________________

____

____

2 ______________________________

____

____

3 ______________________________

____

____

Answer

Write and share ways that Patrice tried to save money. Use the list on page 80.

Answer

1. What did Patrice use to create her budget?

   a. gross pay b. deductions c. net pay

2. What did Patrice use to save money on her grocery bills?

   a. green bags b. coupons c. credit cards

3. What happened when Patrice charged too much on her credit card?

   a. nothing

   b. she got a discount

   c. she had to stop buying lattes

Construction Worker

Diandre works as a construction worker and gets paid weekly. He didn't have a budget and spent all the money he earned. Some weeks he struggled paying his phone bill and rent because he didn't have enough. Finally, Diandre made a budget using his net pay. He budgets weekly for the money he needs for fixed expenses.

Anytown Construction Company
22 Industrial Park
Anytown, USA 53778

Check Number: 8378
Pay Date: 01/25/13

PAY ***Four hundred sixty-four dollars and 74 cents*******************$464.74

Pay to the order of Diandre Jefferson
926 Oak Avenue Apt. 24
Anytown USA 53778

Answer

How much money does Diandre budget each week?

a. $4.64 b. $46.47 c. $464.74

Answer

1. What happened when Diandre didn't have a budget?

   a. he had plenty of money

   b. he struggled paying some of his bills

   c. his parents helped him pay for his bills

2. What are fixed expenses?

   a. bills that stay the same each month

   b. bills that change from month to month

   c. bills that are paid once a year

3. Write one fixed expense Diandre needs to have in his budget.

   ____

Diandre's Budget

Diandre lives with a roommate and shares the cost of his fixed expenses. Shared expenses include rent, Internet, cable TV, heat, electricity, and water. He has to be sure he has money for these expenses. Diandre pays his own cell phone bill, car payment, and car insurance. Why does Diandre need to budget for fixed expenses?

Diandre's Weekly Budget		Income $464.74	
Fixed Expenses	Amount	Variable Expenses	Amount
rent	$137.50	groceries	$45.00
car payment	$50.25	gas/car repairs	$80.00
car insurance	$37.25	entertainment	$30.00
cell phone	$15.00	clothes	$25.00
Internet/cable TV	$5.00	misc.	$26.99
heat	$5.00		
electricity	$6.25		
water	$1.50		

Answer

What is one fixed expense that Diandre must pay?

a. car payment   b. entertainment   c. misc.

Answer

1. What is one fixed expense Diandre shares with his roommate?

   a. rent   b. gas/car repairs   c. groceries

2. What is Diandre's largest variable expense?

   a. groceries   b. gas/car repairs   c. clothes

3. Diandre wanted to buy a pair of jeans for $49.00. Does he have enough money to buy them?

   a. yes   b. no

Searching for the Best Buy

Diandre likes clothes but is careful how he spends his money. He can only budget $25.00 a week for clothes. Diandre watches for sales. He buys a lot of T-shirts because he wears them to work, at home, or when he goes out with friends. When the weather is colder he layers T-shirts under sweaters to keep warm.

Which is the best buy?

Answer

How much money does Diandre have to spend on clothes?

a. $20.00   b. $15.00   c. $25.00

Answer

1. If Diandre bought **2** T-shirts at 50% off how much would they cost?

   a. $10.00   b. $20.00   c. $15.00

2. If Diandre bought **2** T-shirts using the buy-one-get-one free sale how much would they cost?

   a. $10.00   b. $20.00   c. $15.00

3. Which is the better buy?

   a. Sale 50% Off   b. Buy 1 Get 1 Free

Concert Tickets

Diandre's roommate bought concert tickets to see Diandre's favorite band. Diandre has to pay his roommate $60.00 for the ticket. But first he looked at his budget. Daindre decided to use $30.00 for entertainment and $25.00 for clothes to help pay for the ticket. He still needs $5.00 more. He went online to look for coupons to save money.

Online Coupon - Print out

Anytown Gas Station

32 Main Street
Anytown, U.S.A.

10% off
one tank of gas

Limit one coupon per customer

Expires 10/6/15

Answer

What percent will Diandre save using this coupon?

a. 1% b. 10% c. 100%

Answer

1. Diandre filled his gas tank and it cost $60.00. How much money did he save using the coupon? (show work)

2. Does Diandre have enough money to pay for the ticket?

   a. yes b. no

3. Where did Diandre find the coupon?

   a. online b. in a newspaper c. in the mail

A Raise

Diandre got a raise and now earns $25.00 a week more. He decided to start a savings account to cover unexpected costs such as car repairs. Diandre's bank withdraws $15.00 a week from his paycheck and puts the money into a savings account. The bank pays Diandre money, called interest, on the balance in his savings account.

Diandre's Weekly Budget		Income	$489.74
Fixed Expenses	Amount	Variable Expenses	Amount
rent	$137.50	groceries	$45.00
car payment	$50.25	gas/car repairs	$80.00
car insurance	$37.25	entertainment	$30.00
cell phone	$15.00	clothes	$25.00
Internet/cable TV	$5.00	misc.	$26.99
heat	$5.00		
electricity	$6.25		
water	$1.50		
savings	$15.00		

Answer

How should Diandre spend the $10.00 he has left over from his raise?

____

____

Answer

1. If Diandre saves $15.00 a week, how much money will he have saved in a month? (show work)

2. After saving for a month, Diandre had a car repair bill for $40.00. Does he have enough saved to pay for the bill?

   a. yes b. no

3. If you were Diandre, how would you budget the $25.00 a week raise?

____

____

Chapter 4

Monthly Budgets

Vocabulary

1
car payment

money paid for a loan on a car

2
check register

a way of keeping track of a person's money that is in a checking account

3
debt

money usually owed to a bank

4
house payment

money paid for a loan on a house

5
insurance

money paid to protect against theft or damage

6
loan

money borrowed and paid back with interest

Car Mechanic

Jessica

Jessica is a car mechanic. She gets paid once a month. Jessica created a budget so her money would last all month. She has a car loan. Her **car loan** is money she borrowed from her bank to buy her car. Jessica's **car payment** is the money she pays to the bank for the loan.

Jessica's Budget		Monthly Income $2896.00	
Fixed Expenses	**Amount**	**Variable Expenses**	**Amount**
rent (includes heat & water)	$865.00	groceries	$320.00
electricity	$38.00	gas/car repairs	$300.00
Internet	$25.00	entertainment	$200.00
smart phone	$99.00	clothes	$228.50
streaming movies	$10.00	misc.	$285.50
car payment	$210.00		
car insurance	$80.00		
savings	$200.00		
health club	$35.00		

Answer

What is Jessica's largest expense for her car?

a. gas/car repairs b. car payment c. insurance

Answer

1. Jessica's rent also includes:

   a. heat and electricity b. heat and water c. water and electricity

2. Jessica saw a pair of designer shoes in a store that cost $180.00. Did she budget enough money for clothing to buy these shoes?

   a. yes b. no

3. Jessica had $10.00 left over from last month's paycheck. What do you think she should do with the $10.00?

   a. spend it

   b. use it in this month's budget

   c. put it into her savings account

Shopping Online

Jessica found a pair of shoes on the Internet that were $40.00 cheaper than those she saw in a store. There was no tax and the shipping charge was only $4.99. Jessica used her credit card to buy them. She deducted the total amount from her checking account to pay off her credit card bill when it comes.

Anytown Bank Online Check Register

CHECK NUMBER	DATE	TRANSACTION DESCRIPTION	PAYMENT	DEPOSIT	BALANCE
					10.00
				2896.00	2996.00
ETF	7/1	My Landlord	900.00		2096.00
ETF	7/2	Anytown Electricity	84.00		2012.00
TXFR	7/2	Transfer to savings account	150.00		1862.00
ETF	7/2	Internet	25.00		1837.00
ATM	7/2	Cash	20.00		1817.00
ETF	7/3	My Smart Phone	98.00		1739.00
	7/3	Shoes on Internet			

Answer

A **check register** is a way to keep track of money in a checking account. A check register can be done online or by checkbook. What is a check register?

a. a checking account

b. a way to keep track of money in a checking account

c. a savings account

Answer

1. How much did Jessica's shoes cost including shipping? (show work)

2. Write the total cost of the shoes in Jessica's check register on page 98. What is her balance?

3. Look at Jessica's budget to see if she has money left to buy more clothes this month.

a. yes b. no

A Weekend Vacation

Next month Jessica and her best friend are renting a cabin near a lake for the weekend. Jessica needs $375.00. She can take $200.00 out of her savings account but still needs $175.00. To prepare for her vacation, Jessica saved money by cutting her spending on entertainment and clothes. She budgeted $100.00 on entertainment and $153.00 for clothes.

Revised Budget for this month			
Variable Expenses	Amount	Vacation	What's left
entertainment	$200.00	$100.00	$100.00
clothes	$228.50	$75.00	$153.50

Answer

Jessica found a swimsuit on sale for $53.66. Does she have enough money left in her budget to buy it?

a. yes   b. no

Answer

1. What is Jessica's largest expense for her vacation?

   a. cabin   b. groceries   c. gasoline

2. Jessica is bringing $50.00 for miscellaneous expenses. Tell one way she can spend this money.

   ____

   ____

3. Jessica and her friend have to drive a long distance to their rented cabin. Can she save money on gas by sharing the ride with her friend?

   a. yes   b. no

A New Apartment

Jessica found a larger apartment. It's within walking distance of her friends. She can save on gas. There is a free fitness center in the building. Jessica can stop her gym membership. The apartment rents for $1,003. Help Jessica make a new budget to find out if she can afford it.

Jessica's New Budget — Monthly Income $2896

Fixed Expenses	Amount	Variable Expenses	Old Amount	NEW Amount
rent (includes heat & water)	$1003	groceries	$320	
electricity	$42	gas/car repairs	$300	
Internet	$25	entertainment	$200	
smart phone	$99	clothes	$228.50	
streaming movies	$10	misc.	$285.50	
car payment	$210			
car insurance	$80			
savings	$150			
Total Fixed Expenses	$1619			

Answer

Will most of Jessica's fixed expenses stay the same after she moves?

a. yes

b. no

Answer

1. How much money does Jessica have left after she budgets for her fixed expenses? (show work)

2. Write in the amount Jessica can spend each month on her variable expenses. Use the chart on page 102. Remember she cannot budget more money than she earns. Share the budget you've made for Jessica.

3. Can Jessica afford her new apartment?

   a. yes

   b. no

Jessica's Budget

Jessica gets paid once a month. She created a budget to manage her money for the entire month. Jessica didn't want to spend all her money before her next paycheck. When she planned her weekend trip, she adjusted her budget to save for the trip. Before moving, Jessica created a new budget to afford a bigger apartment.

Planning Ahead

saved for a vacation

created a budget for a new apartment

searched online for cheaper shoes

Answer

Do you think Jessica managed her budget well?

a. yes   b. no

Answer

1. Why did Jessica look for sales online?

____

____

2. Instead of saving money for her trip, Jessica could have charged it all on her credit card. Would that have been a better plan?

   a. yes   b. no

3. Write one way you think Jessica managed her budget well.

____

____

____

A Computer Programmer

José works as a computer programmer for a large company. He bought a small townhouse near his job. His budget has money to pay for a **house payment.** That means he pays the bank for his house loan. He also pays house **insurance.** He needs insurance to protect his house against theft or damage.

José's Budget — Monthly Income $4175

Fixed Expenses	Amount	Variable Expenses	Amount
townhouse payment	$856	groceries	$220
house insurance	$100	gas/car repairs	$520
SUV car payment	$596	entertainment	$200
car insurance	$86	clothes	$215
smart phone	$98	credit card	$20
heat	$120	misc.	$567
H.O.A. fees for house	$215		
electricity	$85		
water	$12		
Internet	$30		
satellite TV	$60		
savings	$175		

Answer

People who own townhouses must pay H.O.A. (Home Owners' Association) fees. That means José pays to have someone mow his lawn, remove snow, and trim trees for him. ◯ all of the services José's H.O.A fees cover.

a. mowing the lawn   b. removing snow   c. trimming trees

Answer

1. Does José have a savings account?

   a. yes   b. no

2. José wants to go with a friend to a NBA game. The ticket costs $72.50. He will use the money from his entertainment budget. Does he have enough money to buy the ticket?

   a. yes   b. no

3. Including his insurance and H.O.A. fees, how much money does José spend on his house each month? (show work)

Eating Out

During the week, José rarely cooks. In the morning he buys coffee and a roll. He eats lunch at a sandwich shop. For dinner José orders take-out food to pick up on his way home. He budgets $30.00 a day from his miscellaneous expenses to pay for these meals. Did José stick to his budget today?

Anytown Chinese Food
58 Second St., Anytown, USA 53778

Receipt

ITEM	PRICE
NOODLE BOWL	$7.65
TAX	$.38
TOTAL	$8.03
TIP	$1.00
TOTAL	$9.03

The Coffee Shoppe
135 Main St., Anytown, USA 53778

Receipt

ITEM	PRICE
COFFEE	$3.75
DANISH	$2.35
SUBTOTAL	$6.10
TAX	$.31
TOTAL	$6.41
TIP	$1.00
TOTAL	$7.41

Anytown Sandwiches
200 First St., Anytown, USA 53778

Receipt

ITEM	PRICE
CHICKEN PANINI	$6.99
DIET SODA	$2.35
SUBTOTAL	$9.34
TAX	$.52
TOTAL	$9.86
TIP	$2.00
TOTAL	$11.86

Answer

How much does José budget each day for meals?

a. $30.00 b. $3.00 c. $300.00

Answer

1. At which restaurant did José spend the most money?

   a. Anytown Coffee Shop

   b. Anytown Sandwich Shop

   c. Anytown Chinese Food

2. What was the total amount José spent on meals? (show work)

3. Did José stick to his budget?

   a. yes b. no

Electronic Gadgets

José budgets $200.00 a month for entertainment. He loves electronic gadgets and spends most of this money buying computer games or downloading movies to his laptop or electronic tablet. José also pays for an online game he plays with friends and downloads music and books. He used part of his entertainment money to buy an NBA ticket.

Entertainment Money		
 New computer game **$49.99**	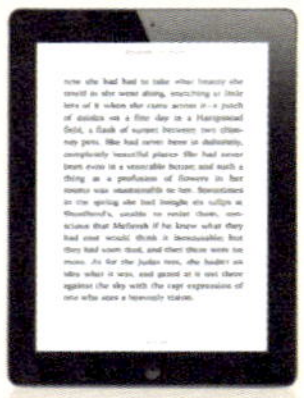 Downloaded book **$9.99**	 Downloaded 4 movies **$59.96**
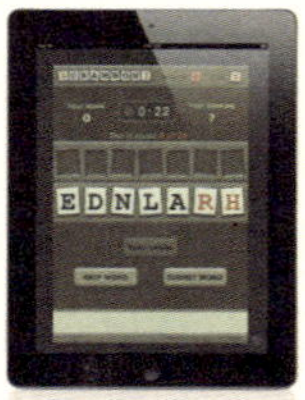 Online computer game **$15.00**	 Downloaded 20 songs **$19.80**	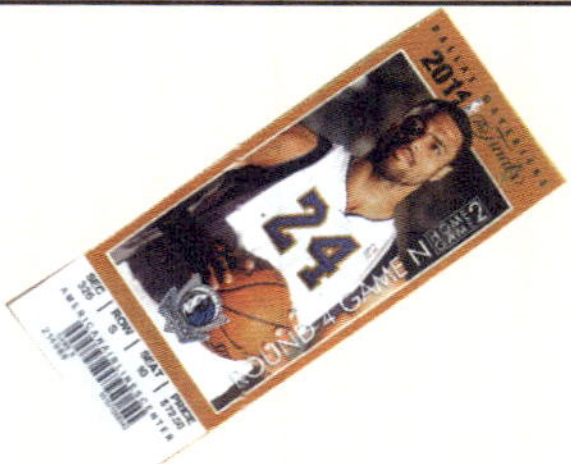 NBA ticket **$72.50**

Answer

How much does José budget for entertainment each month?

a. $20.00　　b. $200.00　　c. $72.50

Answer

1. Write one entertainment expense that José had for this month.

____

2. José spent $227.24 on entertainment. Did he stay within his budget?

   a. yes　　b. no

3. José needs to cut back on other expenses. What do you think he should do?

   a. eat out less　　b. buy fewer clothes　　c. drive less

Paying Down Debt

José is updating his budget. He wants to pay off his car loan in 11 months and have less debt. **Debt** is money he owes the bank for his loan. If José pays $100.00 a month more, his loan will be paid off sooner. José must cut $100.00 from variable expenses and use it to reduce his debt.

José's Budget		Monthly Income $4175	
Fixed Expenses	**Amount**	**Variable Expenses (old amount)**	**NEW Amount**
townhouse payment	$856	groceries ($220)	
house insurance	$100	gas/car repairs ($520)	
New car loan payment	$696	entertainment ($200)	
car insurance	$86	clothes ($215)	
smart phone	$98	credit card ($20)	
heat	$120	misc. ($567)	
H.O.A. fees for house	$215		
electricity	$85		
water	$12		
Internet	$30		
satellite TV	$60		
savings	$175		

Answer

How long will it take José to pay off his car loan using his new budget?

a. 1 month   b. 11 months   c. 11 days

Answer

1. What is debt?

   a. money usually owed to a bank for a loan

   b. a savings account

   c. a checking account

2. Make a new budget for José by changing his variable expenses so he can pay off his car loan. Remember, you have to subtract $100.00 to make sure José doesn't spend more than he earns. Use the chart on page 112 to write your answers.

3. How much extra money will José have after he has paid off his car loan?

   a. $596   b. $696   c. $100

José's Budget

José made a new budget to better spend and save his money. He loves electronic devices and sometimes spends too much. When this happens, he can't eat out as often or buy clothes. José decided to reduce his debt. He paid off his car loan. Now he has more money to spend and save for other things.

Answer

How do you think José should budget his extra money?

a. put some in a savings account

b. spend more on electronics

c. buy a new SUV

Answer

1. What happens when José doesn't stick to his budget?

____

____

2. Write one thing you liked about José's budget.

____

____

3. Write one way you would change José's budget.

____

____

Chapter 5

Loans and Investments

Vocabulary

1

bank CD

a way to invest money in a bank

2

a fine

money a bank charges when a loan is not paid on time

3

401K plan

a way to save for old age

4

investments

ways to earn extra money

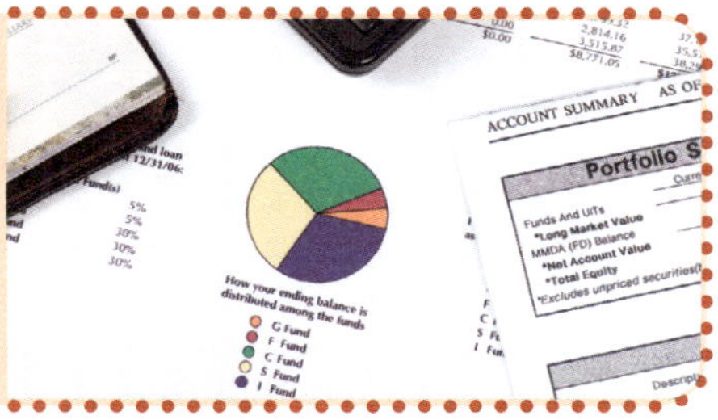

5

profit

money earned from an investment

6

shares of stocks

a person who buys shares owns part of a company

Loans

There are many different kinds of loans. A loan is money a person borrows from a bank. The bank charges interest for loans. Once a month the bank sends a bill stating how much must be paid and when it has to be paid. Loans must be paid on time.

Kinds of Loans

credit cards

car loans

home loans

school loans

ACCOUNT NO. **1234567**	PAYMENT NO. **45**	DATE DUE **5/1**	AMOUNT **$210.00**
		IF PAID AFTER **5/15**	AMOUNT **$240.00**

LATE FEE: $30.00 FINE ADDED TO PAYMENT

Jessica Engleman
15555 E. Washington Ave.
Anytown, U.S.A. 53776

Amount enclosed: two hundred ten and ----------no/100

Send to:
Anytown Bank
601 1st Avenue
Anytown, U.S.A. 53778

Answer

How much does Jessica pay on her car loan each month?

a. $21 b. $210 c. $245

Answer

1. Jessica's car payment for her loan is the same every month. What kind of expense is it?

   a. fixed expense b. variable expense

2. When is Jessica's car loan payment due?

   a. 5/1 b. 5/15

3. What happens if Jessica's car payment is late?

   a. she pays less b. she pays more c. nothing happens

Credit Card Loans

1 Patrice buys something.

2 The store sends the total amount to the bank.

3 The bank pays the store for the total amount that Patrice paid by credit card.

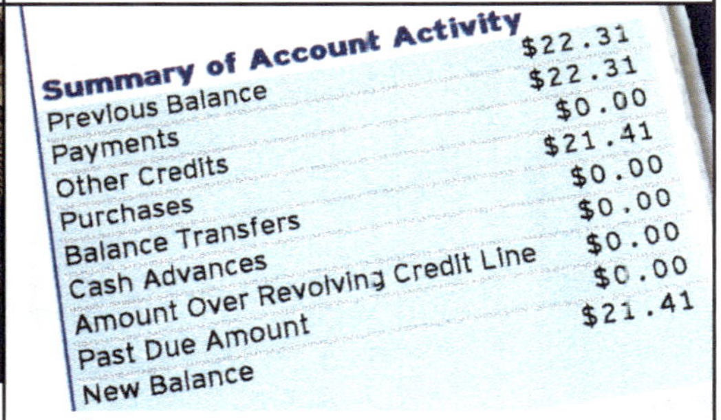

Summary of Account Activity	
Previous Balance	$22.31
Payments	$22.31
Other Credits	$0.00
Purchases	$21.41
Balance Transfers	$0.00
Cash Advances	$0.00
Amount Over Revolving Credit Line	$0.00
Past Due Amount	$0.00
New Balance	$21.41

4 The bank sends Patrice a bill for her purchases.

Jessica only uses her car loan to buy a car. Patrice has a credit card from her bank. Her credit card lets her buy things such as clothes or meals in restaurants. When Patrice uses her credit card, it is called a charge. Her bank pays the amount of the charge to the store and sends Patrice a bill.

Anytown Boutique

8 West Mall Anytown, USA 53778

CASHIER HENRIETTA DATE 7/5/15

Sales Receipt

ITEM	PRICE
SHOES	$59.99
BLOUSE	$23.89
SUBTOTAL	$83.88
TAX	$5.03
TOTAL	$88.91

PAYMENT CREDIT CARD

X *Patrice Novak*

NOVAK, PATRICE

Answer

What was the total amount that Patrice charged?

a. $5.03 b. $83.88 c. $88.91

Answer

1. Why do you think Patrice has a credit card?

   a. to buy a car b. to buy many things c. to save money

2. Who pays the store when Patrice uses her credit card?

   a. Patrice b. a charge c. her bank

3. What does her bank send Patrice when she uses her credit card?

   a. a bill b. a charge c. a discount

What to Know About Loans

There are many kinds of loans. People who have loans usually get them from banks. When a person gets a loan, the bank charges extra money called interest. That means the loan costs more than the amount borrowed. The loan must be paid on time, or the bank charges a fine or extra money for late payments.

Kinds of Loans

A Sample Loan

Jessica wants to buy a new car. It will take her five years to pay off the loan.

Price of the car	$12,500
Interest paid	$2,707
Total cost of loan	$15,207

Answer

What was the price of the car?

a. $12,500   b. $2,707   c. $15,207

Answer

1. The extra money a bank charges for a loan is called:

   a. interest   b. principal   c. borrower

2. How much interest will Jessica pay on her car?

   a. $12,500   b. $2,707   c. $15,207

3. What happens if Jessica pays her loan off in less than 5 years?

   a. she pays a fine

   b. she pays a discount

   c. she pays less interest

Paying Loans

Car loans have a fixed amount to pay each month. Credit card loan payments change from month to month. When a bank sends a credit card bill, the entire amount doesn't have to paid. However, there is a minimum amount that must be paid. Always pay loans on time.

Anytown Bank & Trust Credit

113 S. Main ▪ Anytown, USA 53778

NEW BALANCE	INTEREST CHARGED	MINIMUM PAYMENT	DUE DATE
$225.00	$3.60	$15.00	5/7/14

Credit Limit $2,500 Available Credit $2,275

Account Summary:

Previous Balance	Last Payment	Interest Rate
$125.00	$10.00	16%

Purchases (charges) this month:

Anytown Boutique	$88.91
Sandwich Shoppe	$11.09
TOTAL PURCHASES	$100.00

Answer

What is the minimum payment that must be paid?

a. $10 b. $15 c. $100

Answer

1. Car loans have a:

    a. fixed amount to pay each month

    b. varied amount that changes from month to month

    c. charge no interest

2. Credit card payments have a:

    a. fixed amount to pay each month

    b. varied amount that changes from month to month

    c. charge no interest

3. Loans must be paid:

    a. late

    b. on time

    c. whenever a person wants to pay

Interest and Fines

When José borrowed money for a car loan, the bank charged him interest. Patrice was also charged interest for using her credit card. José and Patrice must pay their bills on time. If a loan is paid late, the bank charges extra money or a **fine** for late payments.

Anytown Bank & Trust Credit

113 S. Main ▪ Anytown, USA 53778

NEW BALANCE	INTEREST CHARGED	MINIMUM PAYMENT	DUE DATE
$225.00	$3.60	$15.00	5/7/14

Credit Limit $2,500 Available Credit $2,275

Account Summary:

Previous Balance	Last Payment	Interest Rate
$125.00	$10.00	16%

Purchases (charges) this month:

Anytown Boutique	$88.91
Sandwich Shoppe	$11.09
TOTAL PURCHASES	$100.00

Answer

A due date means that the amount paid for the bill has to be at the bank by that date. When does the bank have to get Patrice's payment?

____

____

Answer

1. The extra money a bank charges for a loan is called:

   a. interest   b. fine   c. principal

2. If Patrice pays her credit card late, the bank will charge her extra money called:

   a. interest   b. fine   c. principal

3. Tell one thing you know about loans.

____

____

____

What Are Investments?

Investments are ways to earn extra money. Extra money earned is called **profit.** Profits help a person make money. There are many kinds of investments. Depositing money into a savings account or a bank CD are examples. Buying stock in a company or being in a 401K program helps people save money for their future.

savings account
in a bank

certificate of deposit
(bank CD)

stocks
(can go up and down)

401K
a plan to have money when a person is too old to work

Answer

The extra money earned from an investment is called:

a. fines   b. profits   c. savings accounts

Answer

1. Why does a person make an investment?

   a. to get in debt   b. to earn extra money   c. to get a loan

2. One investment a person can make is to get a:

   a. loan   b. credit card   c. bank CD

3. Buying stock is one way to:

   a. get in debt

   b. make an investment

   c. get a loan

Savings Accounts

José decided to open a savings account to earn extra money. He had the bank take out $175.00 each month from his paycheck and put it into his account. José did not spend the money in his savings account. He made a profit of $21.00 from his savings account.

Anytown Bank & Trust

601 1st Ave. ▪ Anytown, USA 53778

SAVINGS ACCOUNT

Account No. 1234

SAVINGS	MONTHLY DEPOSIT	INTEREST RATE	INTEREST PAID	TOTAL SAVINGS
$2,100	**$175.00**	**1%**	**$21.00**	**$2,121**

Answer

How much money did José earn from his savings account?

a. $2.10 b. $21.00 c. $210.00

Answer

1. What is the interest rate José earned on his savings account?

   a. 1% b. 10% c. 1000%

2. How much money did José save each month?

   a. $21.00 b. $210.00 c. $175.00

3. What was José's total savings?

   a. $2,100.00 b. $21.00 c. $2,121.00

Bank CDs

José decided to take some money out of his savings account and put it into a bank CD. A **bank CD** is a way to save money while earning more money than he can in his savings account. To earn a profit, José has to keep his money in a CD for a set time.

Anytown Bank & Trust

601 1st Ave. ▪ Anytown, USA 53778

Certificate of Deposit (CD)

Account No. 1098

DEPOSIT	INTEREST RATE	INTEREST PAID	TOTAL SAVINGS
$2,100.00	**2%**	**$42.00**	**$2,142.00**

Answer

Did José make more profits on his CD than he did on his savings account?

a. yes   b. no

Answer

1. Why did José deposit money into a bank CD?

   a. to keep his money safe

   b. to earn more profit

   c. to pay off debts

2. In order to earn a profit, José must keep his money in a CD how long?

   a. a set time   b. until he wants to use it   c. a month

3. José earned a profit of $21.00 from his savings account. He earned a profit of $42.00 from his CD. How much more money did José earn from his CD? (show work)

Stocks

some electronic companies sell stocks

some stores sell stocks

some restaurants sell stocks

some car companies sell stocks

Companies sell **shares of stocks** to get money to run their businesses. Stock shares sell at different prices. One share can cost \$20.00, \$57.00, \$120.00, or more. A person who buys stock owns part of that company. If the business does well, the value of the stock goes up and the person makes a profit. People buy these shares to make money.

= \$40.00

Answer

How much did each share of stock cost?

a. $40.00 b. $80.00 c. $4.00

Answer

1. Why do companies sell shares of stock?

   a. to get money b. to make a profit c. to save money

2. Why do people buy shares of stock?

   a. to get rid of money b. to make a profit c. to save money

3. If a company loses money, what do you think happens to the people who buy shares of stock?

   a. they lose money b. they make a profit c. they save money

Making a Profit

Patrice's grandmother gave her $4,000. She used it to buy a hundred shares of Terrific Taco stocks. Now Patrice keeps track of the price of her shares. When the price goes up, Patrice can make a profit if she sells her stock. If the price goes down, Patrice will lose money on her stock.

price per share

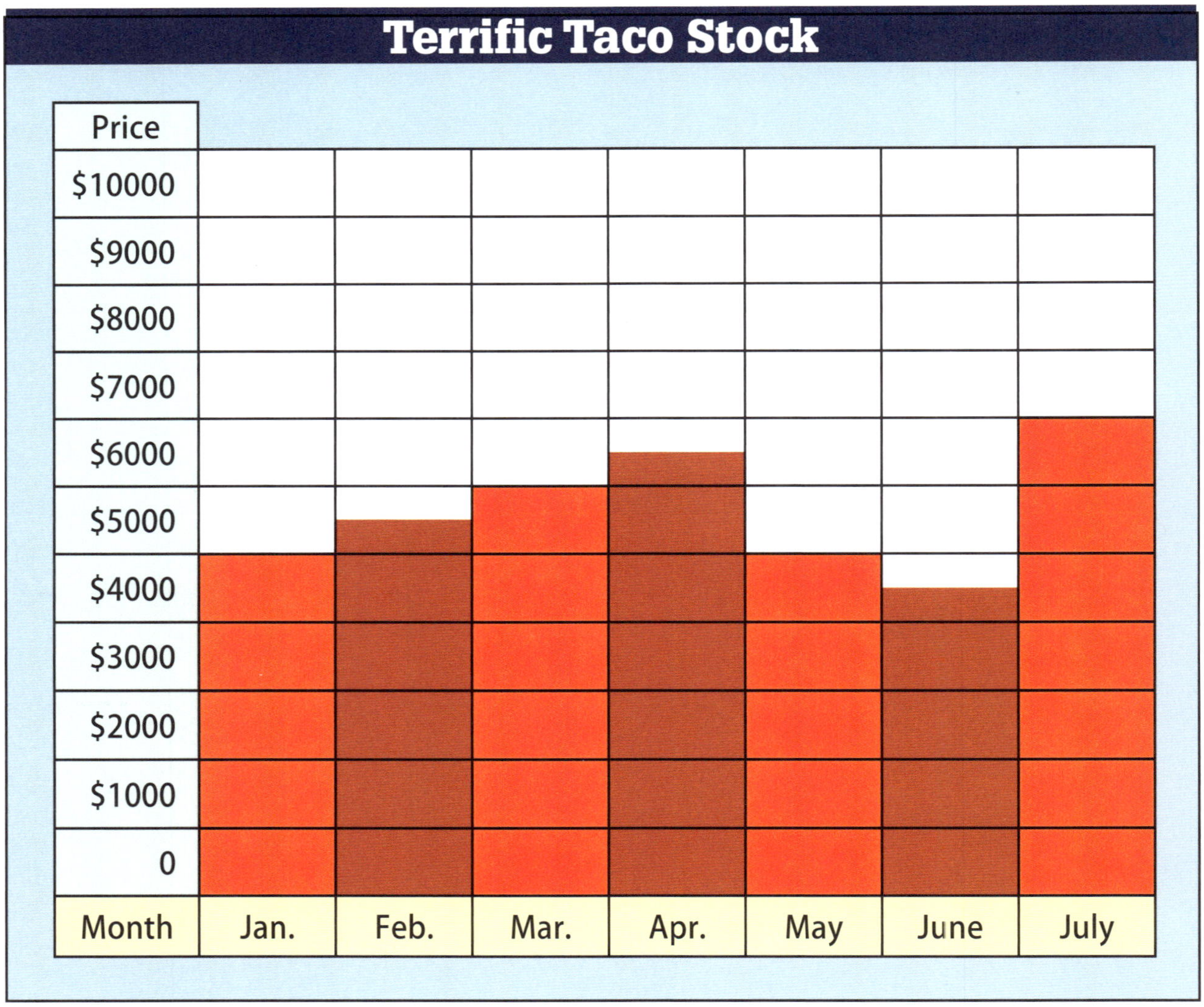

Answer

What was the price per share of stock?

a. $4.00 b. $40.00 c. $400.00

Answer

1. Look at the graph and ◯ the month Patrice should sell her stock to get the best profit.
2. If Patrice sells her shares of stock, how much profit would she make? (show work)
3. Which month would be the worst month for Patrice to sell her stock?

____

Stock Dividends

Terrific Tacos
7258 N. Pacific
Anytown, USA 53778

Check Number: 2788
Pay Date: 01/25/13

Pay to the order of Patrice Novak
Eight hundred eighty dollars and 00 cents******************$880.00

Terry Terrific
authorized signature

The Terrific Taco Company pays dividends to people who bought its stock. Dividends mean that some company profits are paid to people who own stock. Patrice receives a check for her share of the dividends, or company profits.

Answer

How much money did Patrice receive for her shares of stock or dividends?

a. $.80 b. $880 c. $80

Answer

1. A dividend is a check for people who own stock in a company's:

   a. profits b. losses c. products

2. Why did Patrice invest in a company that gave dividends?

   a. to buy things b. to earn a profit c. to spend money

3. What do you think Patrice should do with her profits?

   ____

   ____

401K Plans

José paid his car loan. He now has extra money. Though young, José puts some of his extra money into a retirement plan called a **401K.** That means that José can save money to live on when he is too old to work or wants to retire. He earns interest on the money he invests and will make a profit.

401K Plans

Money + interest = profit

Answer

A 401K plan is a way to save:

a. for a vacation

b. to buy a house

c. to save money to live on when old

Answer

1. Why did José invest in a 401K plan?

   a. to have money for emergencies

   b. to have money to live on when he wants to retire

   c. to have money to buy a new car

2. José's 401K plan earns:

   a. points b. a profit c. credit card

3. Do you think José is smart to start a 401K plan?

   a. yes b. no

Investment Summary

vacations

electronics

money to make more investments

money when people stop working

People make investments to earn extra money, or profits. Investments help people save for the future. People use investment profits to pay for many things. Investments help them save money for when they stop working, or retire. The more money people have in investments, the more profits they earn.

Investment A

Profits: $21.00

Anytown Bank & Trust

601 1st Ave. ▪ Anytown, USA 53778

Certificate of Deposit (CD)

Account No. 1098

DEPOSIT	INTEREST RATE	INTEREST PAID	TOTAL SAVINGS
$2,100	2%	$42.00	$2,142

Investment B

Profits: $880.00

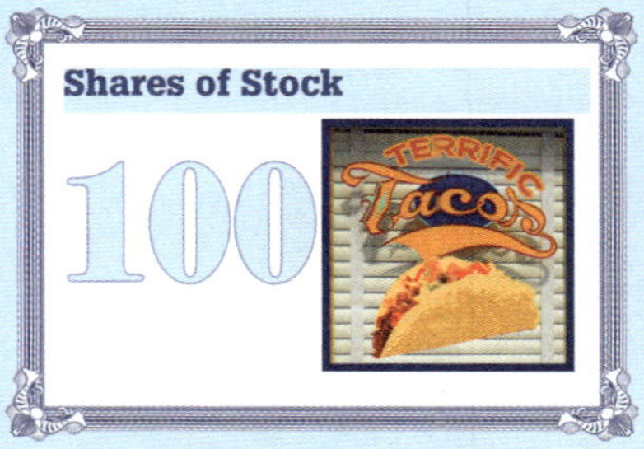

Answer

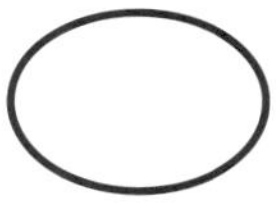 the investment on page 144 that made the most profits.

Answer

1. Investments help people make:

   a. loans b. interest c. profits

2. There are many different kinds of investments.

   a. yes b. no

3. What would you do with profits from an investment?

   a. spend all of it b. spend part of it c. buy more investments

Chapter 6

Putting It All Together

Vocabulary

1
annual income
how much a person earns in one year

2
annual salary
how much a person earns in one year

3
paycheck stub
shows money earned and deductions

4
PIN number
a secret number that the owner of a debit card knows

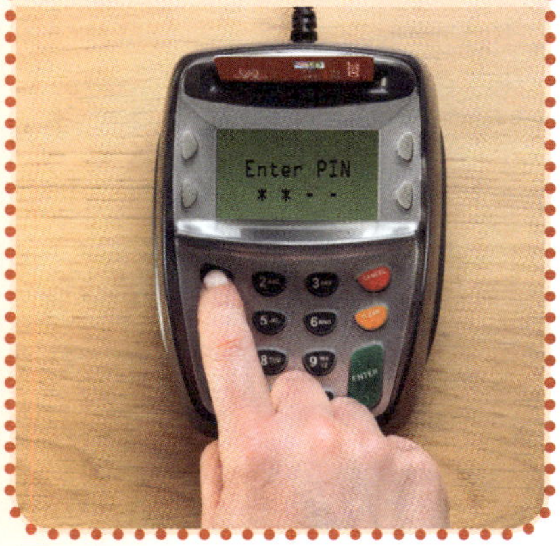

5
taxes
help to pay for roads, schools, etc.

6
utilities
electricity, water, and heat

Putting It All Together

Heilke just graduated from high school. She was hired as a receptionist for a shipping company. Her annual salary is $26,240. An **annual salary** or **annual income** means how much Heilke makes in a year. Heilke moved into an apartment with her friend. She needs to make a budget so that she can keep track of how much she saves and spends.

Annual salary = $26,240

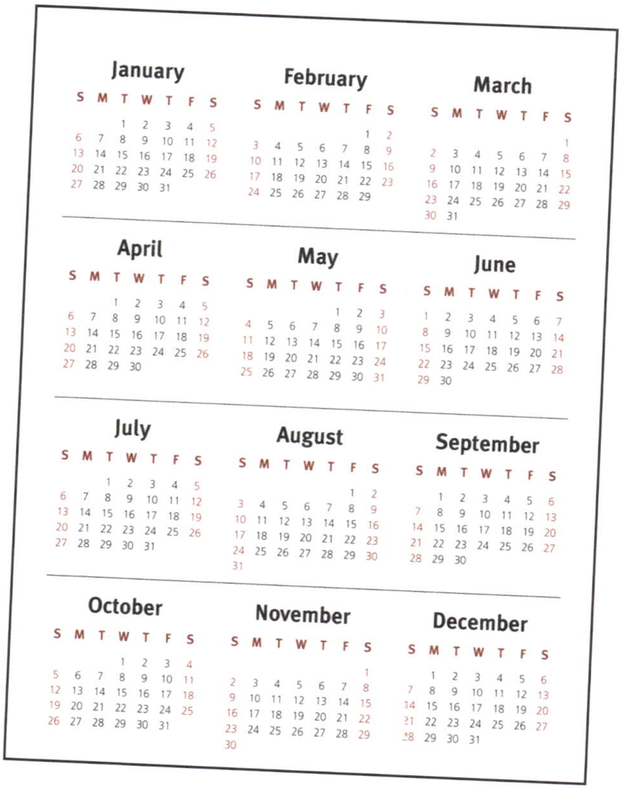

Answer

What is an annual salary?

a. how much a worker earns in a week

b. how much a worker earns in a month

c. how much a worker earns in a year

Answer

1. How much money does Heilke make in a year?

   a. $2,624 b. $26,240 c. $26.24

2. Why does Heilke need to make a budget?

   a. to keep track of how much she spends

   b. to keep track of how much she saves

   c. both a and b

3. Name one expense you think Heilke needs to have in her budget.

   ____

   ____

Reading a Paycheck Stub

Heilke is paid once a month. She reads her paycheck stub so she knows how much money she has to budget with. Heilke's **paycheck stub** shows how much she earns and what her deductions are. Her gross pay is how much money she earns each month. Deductions include taxes, insurance, Social Security, and Medicare. Her net pay is the money Heilke takes home.

Ever-Ready Shipment
7250 Oak Avenue
Anytown, USA 53778

Check Number: 9975
Pay Date: 03/31/13

PAY **One thousand six hundred twenty-two dollars and 5/100*******$1622.05

Pay to the order of Heilke Summerfield
1112 Maple Street
Anytown USA 53778

Heilke Summerfield		**Pay Period 3/1/13-3/31/13**
Earnings	**This Period**	**YTD Amount**
Gross Pay	**$2186.67**	**$6556.00**
Deductions		
Federal Income Tax	$328.00	$934.00
State Income Tax	$109.33	$328.05
FICA - Social Security	$135.57	$406.72
FICA - Medicare	$31.71	$95.12
Health insurance	$69.34	$208.02
Net Pay	**$1512.72**	**$4538.16**

Answer

1. What is Heilke's gross pay?

   a. $2,186.67 b. $6,556 c. $1,512.22

Answer

1. What is Heilke's net pay?

   a. $2,186.67 b. $6,556 c. $1,512.22

2. Heilke needs to create a budget using her:

   a. gross pay b. deductions c. net pay

3. Write one deduction that Heilke pays.

   ____

   ____

What Is a Deduction?

Mr. Sajak, Heilke's boss, explained to her what the deductions are on her paycheck. He told her everyone pays taxes. **Taxes** help to pay for things such as roads or public schools. Other deductions pay for medical expenses and money that retired workers get. One deduction Heilke has is medical insurance. This helps to pay for Heilke's visits to her doctor.

Answer

Heilke's medical insurance deduction helps to pay for what?

a. visits to her doctor

b. roads and school

c. money retired workers can use

Answer

1. Heilke's gross pay is what she earns:

   a. before deductions are made

   b. after deductions are made

   c. by the end of each day

2. What is one deduction Heilke has to pay?

   a. taxes   b. gross pay   c. net pay

3. Taxes help to pay for:

   a. Heilke's visits to her doctor

   b. roads and schools

   c. money retired workers get

Creating a Budget

Heilke has a new job and apartment. She lives with a roommate and pays half of the rent and utilities. Utilities are such things as water, electricity, and heat. To be sure to pay her share, Heilke made a budget. Since she gets paid once a month, her budget covers her expenses and savings for a month.

Heilke's Budget		Monthly Income $1512.22	
Fixed Expenses	Amount	Variable Expenses	Amount
rent	$350.00	groceries	$230.00
electricity	$27.00	clothes	$65.00
heat	$65.00	entertainment	$50.00
water	$7.00	credit card	$10.00
smart phone	$96.00	gas/car repairs	$180.00
Internet/cable TV	$35.00	misc.	$76.03
car payment	$224.19		
car insurance	$67.00		
savings	$30.00		

Answer

Using Heilke's budget on page 157, ◯ a utility that she included in her budget.

Answer

1. What is a budget?

   a. a plan for savings

   b. a plan to pay for utilities

   c. a plan for spending and saving

2. What is Heilke's largest expense?

   a. rent

   b. electricity, water, and heat

   c. gas/car repairs

3. How much does Heilke save each month?

   a. $30.00 b. $3.00 c. $300.00

Fixed Expenses

car payment

apartment

cell phone

The first thing Heilke writes down is how much money she deposits into her checking account each month. Next she makes a list of fixed expenses. Fixed expenses stay the same each month and she pays them first. She must pay bills like rent and utilities before spending money on anything else.

heat

water

Internet/cable

car insurance

savings

Heilke's Budget		Monthly Income $1512.22	
Fixed Expenses	**Amount**	**Variable Expenses**	**Amount**
rent	$350.00	groceries	$230.00
electricity	$27.00	clothes	$65.00
heat	$65.00	entertainment	$50.00
water	$7.00	credit card	$10.00
smart phone	$96.00	gas/car repairs	$180.00
Internet/cable TV	$35.00	misc.	$76.03
car payment	$224.19		
car insurance	$67.00		
savings	$30.00		

Answer

Fixed expenses:

a. stay the same each month

b. change each month

c. aren't important

Answer

1. What is Heilke's largest fixed expense?

   a. rent   b. car payment   c. gas/car repairs

2. Why do you think Heilke has put savings under her fixed expenses?

   ____

   ____

3. How much money does Heilke pay on her car payment and car insurance? (show work)

Variable Expenses

groceries

entertainment

clothes

gas/car repairs

credit card

misc.

Heilke listed variable expenses in her budget. Variable expenses change from month to month. For example, Heilke budgets for gas and repairs for her car. The price of gas and the cost of car repairs can change from month to month. Heilke needs to budget enough for these things so that she has extra money to pay for them if needed.

Heilke's Budget		Monthly Income	$1512.22
Fixed Expenses	Amount	Variable Expenses	Amount
rent	$350.00	groceries	$230.00
electricity	$27.00	clothes	$65.00
heat	$65.00	entertainment	$50.00
water	$7.00	credit card	$10.00
smart phone	$96.00	gas/car repairs	$180.00
Internet/cable TV	$35.00	misc.	$76.03
car payment	$224.19		
car insurance	$67.00		
savings	$30.00		

Answer

Misc. includes money for coffee shops, eating lunch with a friend, or developing pictures for friends to see. What is one thing Heilke can buy with money from the misc. part of her budget?

a. shoes   b. groceries   c. lattes/cosmetics

Answer

1. What is Heilke's smallest variable expense?

   a. groceries   b. credit card   c. entertainment

2. What is one thing Heilke can do to save on groceries?

   a. buy junk food

   b. make a list of groceries she needs

   c. buy only canned food

3. If Heilke has trouble sticking to her budget, what is one expense can she change to not run out of money?

   a. groceries   b. clothes   c. entertainment

Heilke's Checking Account

Heilke opened a checking account to keep track of expenses and savings. She deposits a paycheck into her account each month. Heilke pays most expenses online. The bank withdraws money from her account and pays it to whomever the check is written. Heilke subtracts money from her account each time she pays an expense.

Online Check Register

Account # 2800 800

CHECK NUMBER	DATE	TRANSACTION DESCRIPTION	PAYMENT	DEPOSIT	BALANCE
				$1512.22	$1512.22
728	3/1	My Landlord	$350.00		350.00
					1162.22
729	3/2	Auto Loan Bank	$224.19		224.19
					938. 03
730	3/5	Anytown Insurance	$67.00		67.00
					871.03

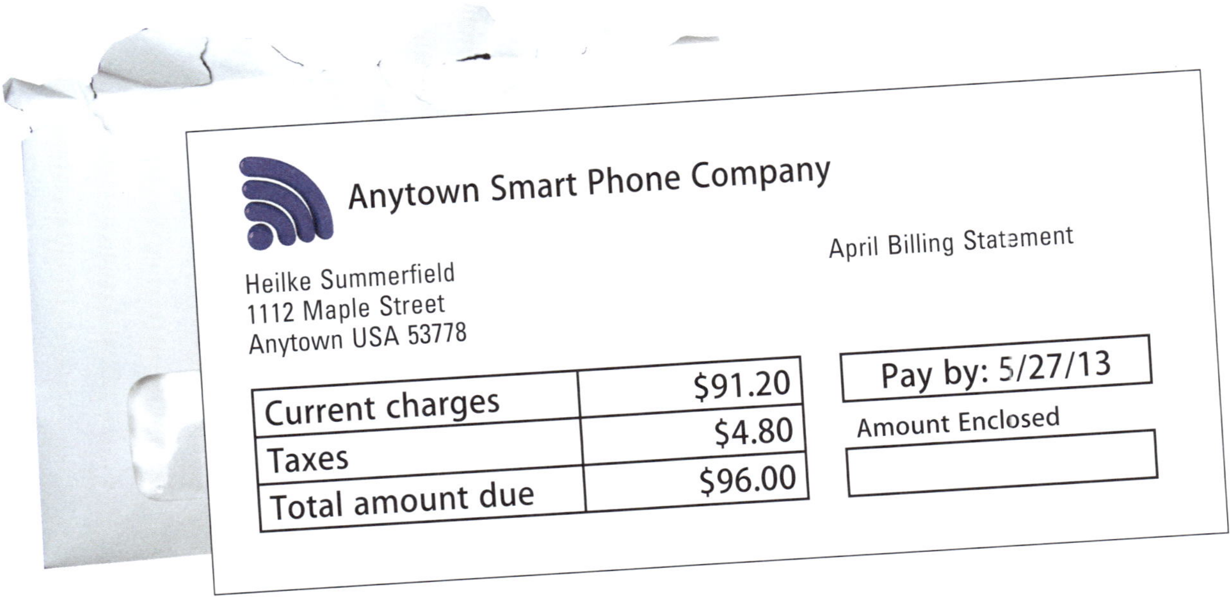
Anytown Smart Phone Company

April Billing Statement

Heilke Summerfield
1112 Maple Street
Anytown USA 53778

Current charges	$91.20
Taxes	$4.80
Total amount due	$96.00

Pay by: 5/27/13

Amount Enclosed

Answer

How much does Heilke have to pay to her smart phone company?

a. $91.20 b. $ 4.80 c. $96.00

Answer

1. How does Heilke use her check register?

   a. to keep track of her expenses

   b. to help her pay online

   c. both a and b

2. When Heilke pays a bill or spends money she must:

   a. subtract the amount from her account

   b. add the amount to her account

   c. keep track of the expense in her head

3. Heilke has $871.03 in her checking account. How much money does she have left after she pays $96 for her cell phone bill? (show work)

Heilke's Savings Account

At the beginning of each month, the bank takes money out of Heilke's checking account and puts it into a savings account. The bank pays Heilke money to keep a savings account. The money the bank pays is called interest. That means the bank pays Heilke to invest in a savings account where Heilke can earn extra money.

Heilke's Savings Account

monthly deposit

Anytown Bank & Trust

601 1st Ave. ▪ Anytown, USA 53778

SAVINGS ACCOUNT

Heilke Summerfield
1112 Maple Street
Anytown USA 53778

Account No. 101010

SAVINGS	MONTHLY DEPOSIT	INTEREST RATE	INTEREST PAID	TOTAL SAVINGS
$360	**$30.00**	**1%**	**$3.60**	**$363.60**

Answer

How much money did Heilke save in one year?

a. $360.00 b. $3.60 c. $363.60

Answer

1. How much money each month did Heilke save?

   a. $3.00 b. $30.00 c. $300.00

2. How much extra money or interest did Heilke earn in one year?

   a. $36.00 b. $3.60 c. $360.00

3. Write one reason why Heilke has a savings account.

   ____

   ____

Deposits and Withdrawals

SAVINGS DEPOSIT

Date 6/15/13

Name Heilke Summerfield

Account Number 101010

Heilke Summerfield
AUTHORIZED SIGNATURE

Anytown Bank & Trust
601 1st. Ave.
Anytown, USA 53778

CASH	100.00
CHECKS	15.00
(list)	
Subtotal	115.00
Less Cash	10.00
TOTAL	105.00

When Heilke deposits money into her checking or savings account, she is putting money into the account. When she gets birthday or holiday money from her family, she puts it into her account. When Heilke writes a check from her checking account or takes money out of her savings account, she is making a withdrawal.

SAVINGS WITHDRAWAL

Date 7/24/13

Name (PRINT) Heilke Summerfield

Heilke Summerfield
AUTHORIZED SIGNATUR

SAVINGS WITHDRAWAL

AMOUNT WITHDRAWN
$100.00

Anytown Bank & Trust
601 1st. Ave.
Anytown, USA 53778

101010

Answer

How much money did Helke withdraw or take out of her savings account?

a. $1.00 b. $10.00 c. $100.00

Answer

1. How much money did Heilke deposit into her account?

   a. $105.00 b. $10.50 c. $1.50

2. What is it called each time Heilke puts money into one of her bank accounts?

   a. deposit b. withdrawal

3. Heilke wrote a check for her weekly groceries. Did she withdraw money from her checking account to pay for the groceries?

   a. yes b. no

Debit Cards

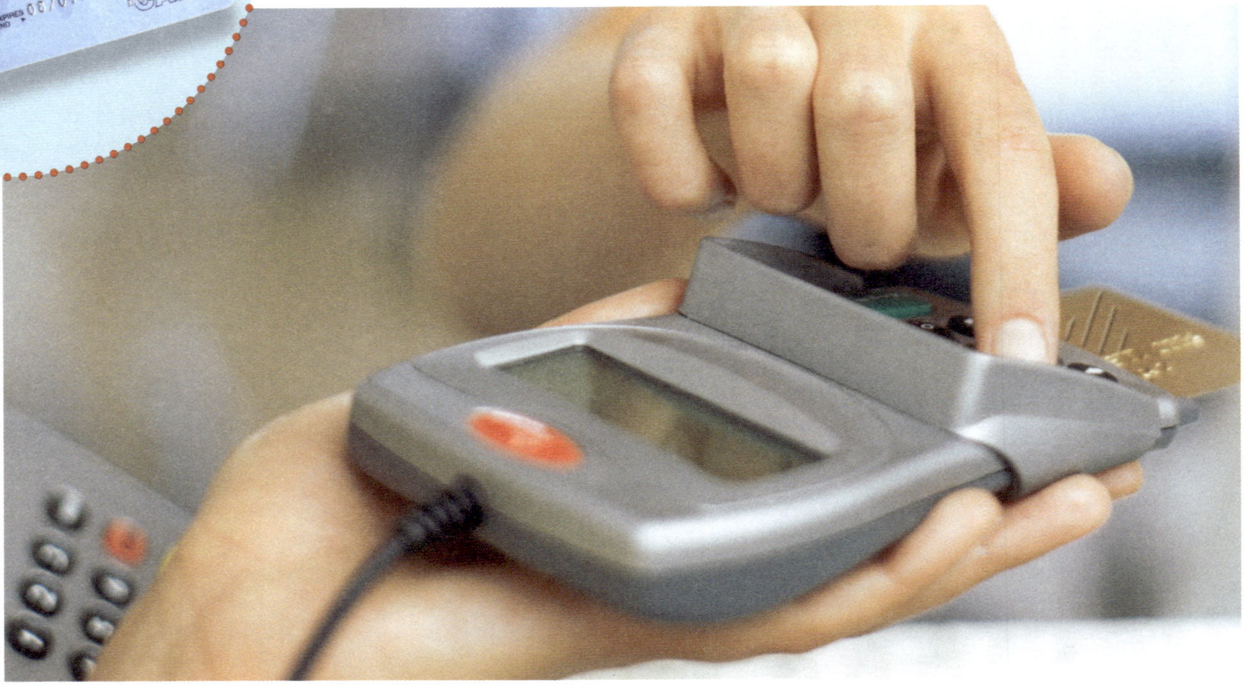

Heilke has a debit card that acts like a check. Her card lets her withdraw money from her checking account to buy things. Heilke keeps the receipts for things she buys. She subtracts the amount she spends in her check register. Her debit card has a PIN number. The **PIN number** is a secret number that only Heilke can use.

Anytown Salon

315 First St., Anytown, USA 53778

STYLIST KERRY

Receipt

ITEM	PRICE
HAIRCUT	$30.00
TAX	$1.50
TOTAL	$31.50
TIP	$6.00
TOTAL	$37.50

PAYMENT DEBIT CARD

X *Heilke Summerfield*

HEILKE SUMMERFIELD

Answer

How much money does Heilke have to subtract from her account?

a. $30.00 b. $31.50 c. $37.50

Answer

1. A debit card acts like a:

   a. savings account b. check c. credit card

2. What does Heilke have to keep in order to subtract her purchase in her check register?

   a. receipt b. PIN number c. debit card

3. What is a PIN number?

   a. a credit card number

   b. a secret number

   c. a house number

Credit Cards

Heilke has a credit card. Heilke can pay for things with it when she doesn't have enough money or orders online. When Heilke uses her credit card, she borrows money from a bank. She has to be careful not to use her credit card too much. The more she charges on her card, the larger her monthly payment.

Anytown Bank & Trust Credit

113 S. Main ▪ Anytown, USA 53778

NEW BALANCE	INTEREST CHARGED	MINIMUM PAYMENT	DUE DATE
$300.00	$3.50	$15.00	3/27/14

Credit Limit $1,000　　Available Credit $700

Account Summary:

Previous Balance	Last Payment	Interest Rate
$250.00	$15.00	14%

Purchases this month:

02/19 Susie's Sandwich Shoppe	$11.50
02/25 Kathy's Coats	$53.50
TOTAL PURCHASES	$65.00

Answer

What is the total amount of new purchases Heilke has charged on her credit card?

a. $300 b. $15 c. $65

Answer

1. Why does Heilke have a credit card?

   a. to pay for things on online

   b. to pay for her car

   c. to get a larger monthly payment

2. What is the minimum amount Heilke has to pay this month?

   a. $300 b. $3.50 c. $15.00

3. What happens when Heilke charges a lot of things?

   a. her monthly payment is larger

   b. her monthly payment is smaller

Heilke's Loan

Heilke bought a car. She applied for a loan from her bank. A loan means the bank lent her money to buy her car. Heilke has to pay the bank back for the loan with interest. Interest is the charge that Heilke's bank made on the money she bought her car with.

Anytown Bank & Trust Auto Loans

113 S. Main ▪ Anytown, USA 53778

Heilke Summerfield
1112 Maple Street
Anytown USA 53778

Account Summary:

NUMBER OF PAYMENTS	INTEREST RATE	AMOUNT BORROWED	MONTHLY PAYMENTS	TOTAL PAID	INTEREST PAID
60	**9%**	**$10,800**	**$224.19**	**$13,451.40**	**$2,651.40**

Answer

How many payments must Heilke make to pay her car loan?

a. 6 b. 90 c. 60

Answer

1. How much is Heilke's monthly payment?

   a. $224.19 b. $2 6.51 c. $22. 41

2. What is the total amount Heilke will pay for her car?

   a. $10,800 b. $224.19 c. $13,451.40

3. Will Heilke pay less interest if she pays off her loan in less than 60 payments?

   a. yes b. no

Heilke's Investment

Heilke's raise

money from Heilke's savings account

After Heilke worked for six months, she got a raise. She decided to put the extra money into an investment. Using money from her savings account and extra money from her paycheck, Heilke started a bank CD. By keeping the money in a CD, she makes a bigger profit and earns more money.

Anytown Bank & Trust

601 1st Ave. ▪ Anytown, USA 53778

Certificate of Deposit (CD)

Account No. 97777

DEPOSIT	INTEREST RATE	INTEREST PAID	TOTAL EARNED
$250.00	**2%**	**$5.00**	**$255.00**

Answer

How much money did Heilke have to start her CD?

a. $200.00 b. $50.00 c. $250.00

Answer

1. Why did Heilke invest in a Bank CD?

   a. to earn extra money

   b. to make a profit

   c. both a and b

2. The interest rate on Heilke's bank CD is 2%. What was the total amount Heilke earned in one year?

   a. $200.00 b. $5.00 c. $255.00

3. Heilke got a raise and now has $50.00 a month more to spend or save. How do you think she should budget her money?

   ____

   ____

Heilke's Summary

Heilke gets paid once a month and created a budget to cover monthly expenses and savings. Heilke read her paycheck stub to see how much money she had after deductions. She was careful not to take out too many loans. Heilke made an investment after she got a raise.

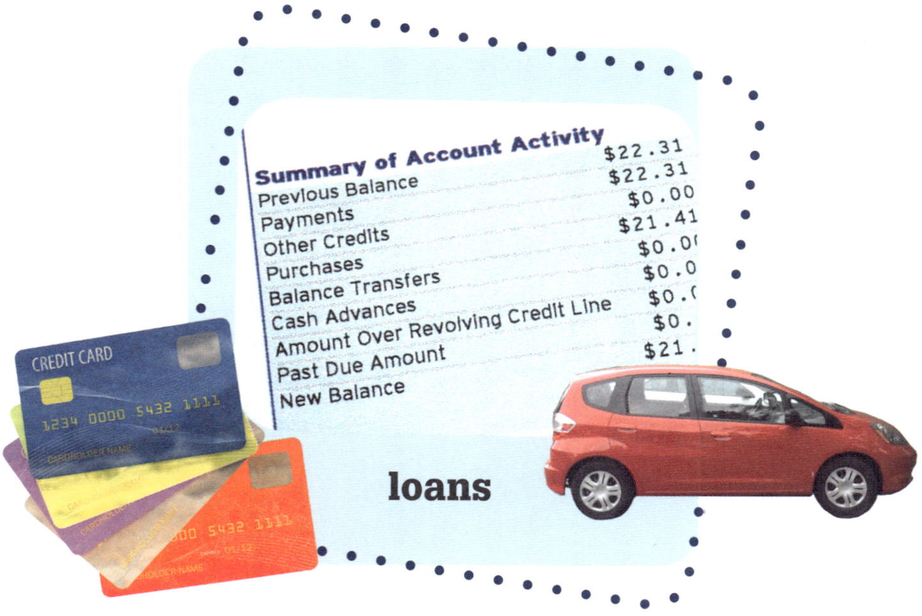

Answer

What is a loan?

a. money paid to protect against theft

b. money borrowed to buy something

c. extra money earned

Answer

1. Why did Heilke make a budget?

____

____

2. Heilke was careful not to have too many:

a. investments b. loans c. savings

3. Name one thing you liked about Heilke's budget.

____

____